International Poultr

Indian Runner Ducks

POULTRY BOOKS by Dr Joseph Batty

The Ancona Fowl   Andalusian Fowl
Artificial Incubation & Rearing
Bantams -- A Concise Guide;   Bantam Management
Bantams & Small Poultry; Brahma & Cochin Poultry
Breeds of Poultry & Their Characteristics
Call Ducks: Concise Poultry Colour Guide
Cockfighting: Poems & Prints
Domesticated Ducks & Geese
Duck Breed Books -- Aylesbury, Orpington, Indian Runners, etc.
Frizzle Fowl ; Garden Poultry Keeping
Hamburgh Poultry Breeds; International Poultry Standards
Japanese Bantams: Japanese Long Tailed Fowl
Jersey Giant Poultry
Khaki Campbell Ducks & the Campbells of Uley
Keeping Jungle Fowl ; Langshan Fowls
The Malay Fowl: Marans
Marsh Daisy Fowl   Minorca Fowl
Naked Neck Poultry
Natural Incubation & Rearing ; New Hampshire Reds
Natural Poultry Keeping:
Old English Game Bantams
Old English Game Colour Guide
The Orloff Fowl
Orpington Fowl (with Will Burdett)
Ostrich Farming: Polish Poultry Breeds
Plymouth Rock Poultry
Poultry Ailments: Practical Poultry Keeping
Poultry Characteristics—Tails
The Poultry Colour Guide
Poultry Foods & Feeding
Poultry for Beginners
Poultry Shows & Showing
Races of Domestic Poultry, Sir Edward Brown
Revised Edition Joseph Batty
Rhode Island Red Fowl
Rosecomb Bantams
Scottish Poultry Breeds
Sebright Bantams; Sicilian Poultry Breeds
The Silkie Fowl;   Spanish Fowl
Sultans; Sumatra Game Fowl;
Sussex & Dorking Fowl
True Bantams
Understanding Modern Game
(with James Bleazard)
Understanding Indian Game
(with Ken Hawkey)
Understanding Old English Game; Welsummer Fowl : Wyandotte Poultry

Indian Runner Ducks

J Barnes

Beech Publishing House
Station Yard
Elsted Marsh
Midhurst
West Sussex GU29 OJT

© Joseph Batty, 2007

This book is copyright and may not be reproduced or copied in any way without the express permission of the publishers in writing.

ISBN 978-1-85736-224-4

First published 1999
New Impression 2007

British Library Cataloguing-in-Publication Data
A catalogue record for this book is available from the British Library.

Beech Publishing House
Station Yard
Elsted Marsh
Midhurst
West Sussex GU29 OJT

CONTENTS

1 Early Origins	1
2 Relatively Modern Progress	15
3 The *Runner Duck Standards*	21
4 Colour Descriptions	31
5 General Rules	
Duck Management	49

PREFACE

Ducks are fascinating birds and none more than the breed known as the Indian Runner Duck. It was given the pre-fix 'Indian' because it was believed to have originated in India, which turned out to be untrue. But to make matters worse a person who did know the true source took the secret to the grave, although, later, a friend revealed that they came from Indonesia, and more specifically, from the Islands of Lombok and Bali.

Whilst the true source was still being debated a Dutch man claimed they came from Holland and there was much controversy, many believing his claims. Fortunately, he was proved wrong and the record put straight.

Indian Runners walk upright without the traditional waddle of the normal domesticated ducks. They are fine layers of white eggs and their productivity is remarkable.

They have been used to cross with other ducks to improve egg production and the main element in Khaki Campbell Ducks is certainly the Indian Runner. The Campbells were developed by Mrs A Campbell about 100 years ago and, quaintly, the colour was a tribute to the those fighting for Queen and country in the Boer war, wearing khaki uniforms.

J Barnes September, 1999

COLOURS IN RUNNER DUCKS

The colours continue to multiply, which is a great pity because the Fancy is not numerically strong enough to maintain the range of colours now being attempted. Far better to concentrate on the few established colours and achieve perfection in those.

White, and Fawn & White Runner Ducks

The *White* is too horizontal and does not comply with the standards.
The *Fawns* are a light fawn colour and therefore quite acceptable, but their bodies are not slim enough for show requirements; they should also be more upright.

Indian Runner Ducks
(Painted by Paul Chapman)
Colours do vary to some extent.

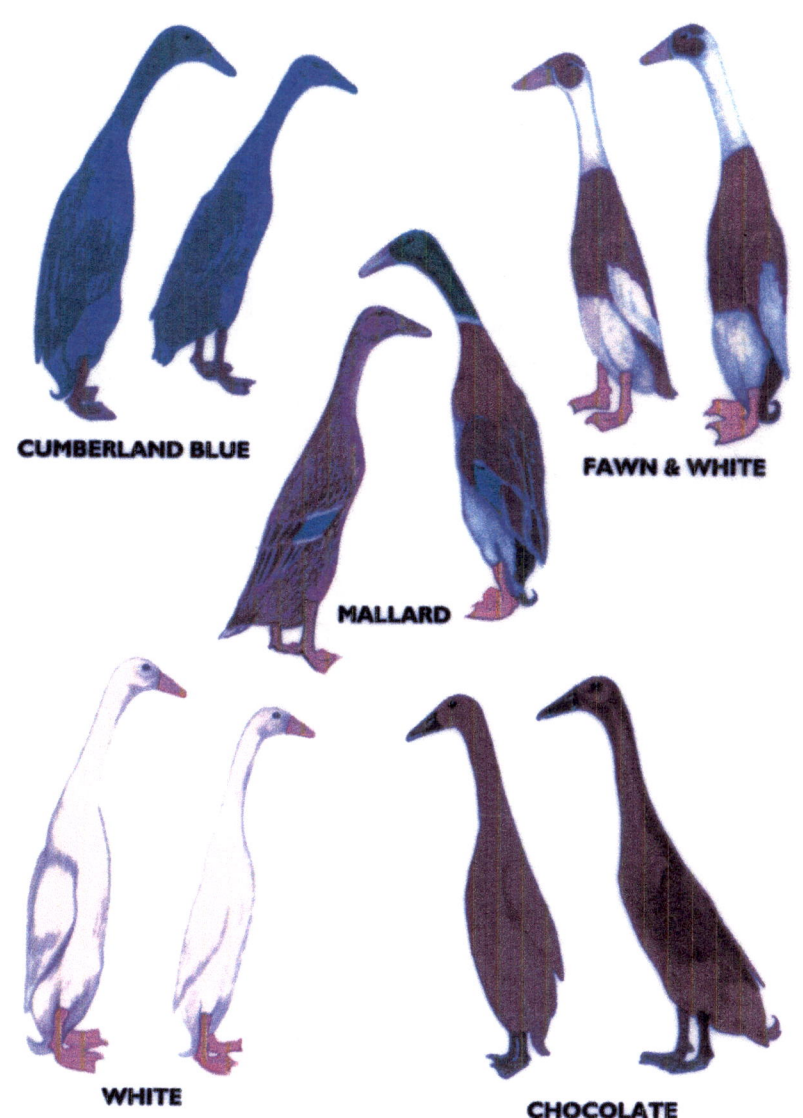

Indian Runner Ducks
(Painted by Paul Chapman)
Colours do vary to some extent.

Runner Ducks

From a painting by A J Simpson

These are not as bottle shaped (rounded) as modern winners, but are fairly representative. Interestingly, the bird on the Left was called a 'Fawn', but is more like what is now termed a Mallard (Grey USA). The other two are White and Fawn & White, but even there the fawn is not quite right based on the standard (see text). Fawn should be an even gingery red colour.

1

EARLY ORIGINS

Fawn & White Indian Runners
The drake, with the curly feathers in its tail, is on the right.

ORIGINS
Not from India*

Although the breed is known as the Indian Runner, it did not originate in India, although for many years this was believed to be the case. In fact, because of the mystery surrounding it, and the deliberate withholding of information, the position remained vague. This went on until 1927 when light was shed on the origins.

Briefly, it arose from birds being brought to this country by a ship's captain, trading *in the Indies*, a vague term in those days. The first import occurred about 1850. They were light in build and they were characterized by an unusual walk or run, inasmuch it was straight-forward and free from waddle. The carriage was erect; the head and bill were unique, wedge shaped, the eye socket high in the skull, the neck thin and racy.

Some of these birds were taken to Cumberland,** where they were crossed with the duck of that county; the remainder were sent to Dumfrieshire, Scotland, where they were bred in the original form and colour—fawn, but mallard blood being introduced from time to time.

---

* The main sources for this chapter are a paper given at the International Poultry Congress, 1930 presented by Miss M Chisholm.

** Some records state that it was the Penguin duck which was the first import and when crossed became the Runner Duck (*The Encyclopedia of Poultry,* J T Brown), but this is incorrect.

The early history is said to have been recorded by Mr J Donald of Wigton in a book entitled *History & Description of the Indian Runner Duck,* *1890.

He believed that they had originated in India and the first imports were around 1840. Once they became popular the indiscriminate crossing took place and they lost many of their original features. In fact, only his small book was able to provide a standard of sorts against which the breed could be stabilized.

The First Show for Runners

The first major stride was in 1896 when a Miss Wilson-Wilson got together a class at Kendal Show which consisted of 22 pairs. This was followed by others and thus they became popular.

Harrison Weir** confirms the developments taking place and mentions the role played by Mr Henry Digby who, with Miss Wilson-Wilson endeavoured to have the breed recognized by the Water-fowl Club so a standard could be issued.

The claim for very high egg production was confirmed by some and queried by others. Undoubtedly the top strains do exceed 240 eggs or more, but there were many that laid much lower figures, no doubt due to the crossing. The relative small size does not make an excellent table duck, but it does have firm flesh, which tends to be 'gamey' in flavour.

---

*Cited by S H Lewer in *Wright's Book of Poultry*, new edition, but we have not seen a copy.
** *Our Poultry*, 1903.

The excellent laying capabilities of the Khaki Campbell breed owe a great deal to the Indian Runner. It was from this breed that Mrs A Campbell of Uley, crossing Runners with the Rouen, and then selecting the top layers, produced the utility layers.

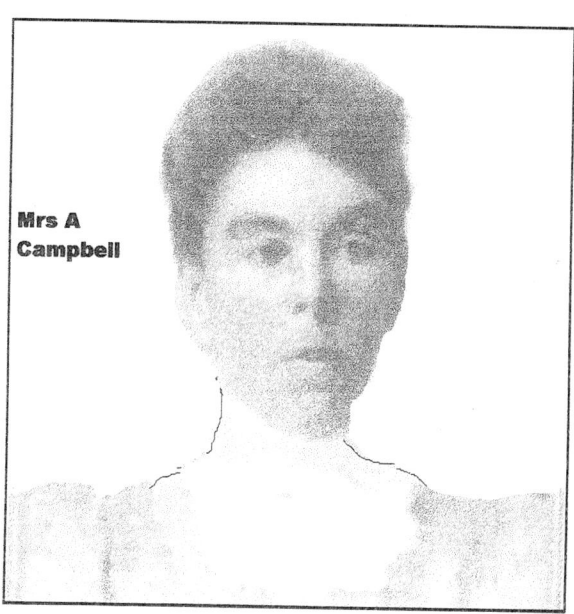

Mrs A Campbell

Mrs A Campbell who in 1901 developed the Khaki Campbell from the crossing of a Rouen drake with a Fawn and White Runner, then crossed the progeny with a Mallard. She became famous for her work with poultry.

INDIAN RUNNER & BALI DUCKS

Khaki Campbell Drake
The breed developed by Mrs A Campbell from crossing Indian Runners, Rouen and Mallard to produce a top layer as well as a reasonable table bird.

INDIAN RUNNER & BALI DUCKS

THE USA

In the USA the Indian Runner was established around 1890, but the type being kept at the beginning of the 20th century were a heavier, more utility type bird. According to a reliable source* the Indian Runner was 'aptly termed the Leghorn of the duck family'. It was known that they ran around and foraged for a great deal of their food, but were light in weight and non-sitters.

Drawing of Indian Runners in the USA, 1918

Note the relatively thick bodies and lack of upright stance. These are certainly of the utility type with a build to produce plentiful eggs and reasonable body weight. These are quite different from the present-day standard bred birds, which are not thick in body.

---

* *Commercial Poultry Raising*, H Armstrong Roberts, 1923,

COLOURS IN THE USA

The colours found in the USA are similar to the English version and include:

1. Fawn & White	(1898)
2. White	(1914)
3. Pencilled	(1914)
4. Black	(1977)
5. Buff	(1977)
6. Cumberland Blue	(1977)
7. Chocolate	(1977)
8. Grey	(1977)

The date shown is the year when the variety was introduced into the American Standards. In the USA the breed is classified as a Light breed with weights of 4.50lb. for mature drakes and 4lb. for mature ducks, with younger birds being lighter.

Exhibition birds must be quite slim and as straight upright as possible. They must be cylindrical in shape, without pronounced shoulders or very rounded bodies which are not in line with the streamlined shape which is essential.

The standards give the shape as being similar to an old type soda bottle, but this term is not understood any more because such bottles are no longer around. What should be remembered is that the body should be almost perpendicular, ie, almost at right angles to horizontal. The older books stipulated 65 to 70 degrees*, but many of the best specimens exceed this figure.

---

* See for example, *The Encyclopedia of Poultry*, **J T Brown**.

A CONTROVERSY

The breed turned out to be excellent layers and were very popular, a boom occurring in the 1890s when they were in great demand.

As a result of the sharp rise in interest more information was desired on the exact origins, because it was felt that there may be better birds at source. However, the mystery remained and this led to bitter controversy.

In the boom of the 1890s birds were exported to all parts of the world, including America, and British fanciers pushed the breed for all they were worth, lauding its praises and claiming for it an Anglo-Indian origin.

A Critic of the Claims

The source was disputed. In 1900 Louis Vander Snickt, a native of Belgium and one of the foremost European authorities on poultry, launched an attack against the British fancier, disclaiming the Indian origin of the bird and suggesting that it was no new breed but had been obtained from no further away than the Low Countries, where ducks of this type had been kept for centuries.

Other claims were made for the story that the Indian Runners had been imported into Norfolk from Holland and then spread into England. Although it is possible that there was some truth in the claim, this did not admit that Holland was the *origi-*

---

* Poultry Breeding & Production, Vol. 3, 1929.

nal source. Most likely, it was argued, the Dutch birds came from Asia.

The American Support for Vander Snickt

Apparently many in America whole-heartedly accepted it, and, in his book *Growing Ducks and Geese,* J H Robinson, of the " Reliable Poultry Journal," Ohio, in a long chapter on the Indian Runner Duck, very definitely declared himself for Mr. Vander Snickt.

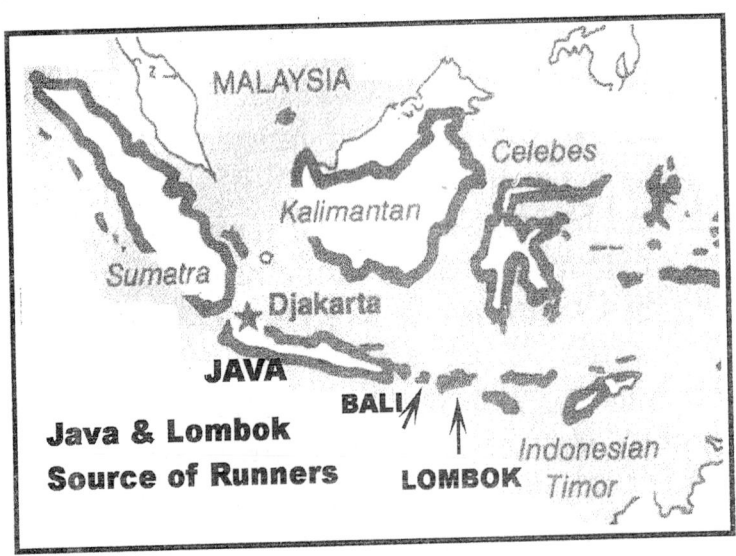

True Source of the Indian Runner Ducks (Lombok)
Ducks with upright carriage also came from Bali and other islands nearby. The Malayan Runner has been in existence for more than 2,000 years on the island of Java.

INDIAN RUNNER & BALI DUCKS

Meanwhile British fanciers stuck to their beliefs. For some time every effort had been made to trace the bird in India, but no evidence was forthcoming that any such bird was, or ever had been, indigenous to that country. This was discouraging, but not for one moment was the Dutch origin accepted.

Unfortunately for the British fancier Mr Walton did not wish to give away the secret of the country of origin of his birds, so that no answer could be given to our critics, and Mr. Vander Snickt's opinion still held good, although his claims could not be verified.

Only after Mr Walton died was the secret revealed. In 1927 a Mr. Matthew Smith, close friend and ally of the late Mr. Walton, and secretary of the Indian Runner Duck Club of Great Britain, gave his permission for all the facts to be made known, and a book, **The History of the Indian Runner Duck**, by Dr. J. C. Coutts, was published by **The Feathered World**, London. In this everything was revealed and documentary evidence given, both in regard to Mr. Walton's and later importations, and that the birds were found in fairly large numbers in Java and in small numbers in Lombok, both islands of the Malay archipelago. It formed a very complete reply to all our critics and it proved the origin of this breed of ducks.

It seems that imports to Britain probably came from Lombok, which, as noted, is a small island off Java. However, as shown by the map, these unusual ducks are quite popular, including Bali Ducks.

INDIAN RUNNER & BALI DUCKS

The Crested Bali Duck.
This appears to be a close relation of the Indian Runner, but why they developed separately is not clear. Why the two types of duck, closely related, should develop along different lines, is a puzzle. Possibly it was following the same kind of separation of the species as occurred on the Galapagos Islands as discovered by Charles Darwin in 1835.

The Malayan Runner, the prototype of the original importation and the bird which gave the name Indian Runner to the breed, is known to have existed 2,000 years ago in the island of Java. Carvings of these ducks with an attendant in charge are to be seen in the Hindu temple of Boeroe Boeddha, which bears indications of the worship of Buddha and Siva. It is a lightly built duck, eminently suited to the tropical swamps in which it thrives, and there is no doubt that it has the ability to pass on persistency in egg production. The actual evolutionary origin of these birds is a questionable point, as they differ in practically every external feature from all other waterfowl. Do they come from the Mallard as other breeds?

To anyone conversant with the map of the Malay Archipelago it will be seen that, in between the islands of Java and Lombok, lies the island of Bali. To further complicate the study of this most interesting duck we find on Bali Island a white crested bird, bearing in the main the characteristics of the Fawn Indian Runner Duck.

Why should this be? By what strange process have these birds been evolved and how is it that the Bali duck, obviously from the same source as the Runner, achieved not only white plumage and a white egg, but a crest as well? Charles Darwin maintained that owing to the presence of the curled tail feathers in the drake the strong probability was that they had sprung from the common wild duck, ie, Mallard. If so, no explanation can be found to account for the amazing change which has taken place in the structural formation nor the presence of the white crested duck of Bali.

Undoubtedly, the Runner duck has done more to improve laying capabilities than any other breed. It has been crossed with other breeds all to improve the laying and the main breed which excels for egg production, the Khaki Campbell, owes its fine reputation from the correct blending of the breeds involved, explained earlier.

2
RELATIVELY MODERN PROGRESS

Utility Runner Type
which offered body weight as well as laying abilities.

NOTE: The original crosses for the Khaki Campbell showed similar tendencies before being refined by selection.

DEVELOPMENT

Early references to the breed are confusing. In one early book the breed is referred to as the Penguin Duck*. It was stated:

> *This very extra-ordinary looking duck is characterized by greater length of femora, or upper bones of the feet, whilst the tibia remains unchanged. In consequence of this peculiarity of structure, the duck, in walking, is obliged to assume an erect attitude, like that of the Penguin.* *

Because of all the uncertainty, the development of the breed has fallen into two distinct categories:

1. *Exhibition or Malayan-type of Runner, and*
2. *Utility Indian Runner, which gained a world-wide reputation as a layer.*

Moreover, it seems that in that the two birds are not identically the same breed owing to the fact that the Utility Runner is in truth a Runner-Cumberland Duck cross, which occurred with the original imports, which may have been the so-called Penguin ducks.

---

* *The Poultry Book*, 2nd ed. W B Tegetmeier, 1873. At this stage it appears little was known of the laying capabilities because the author stated that the breed had little to offer which distinguished from other ducks. At this stage there was certainly confusion and the Poultry Club standards of 1926 show the Penguin duck to be in excess of 7lb or more, whereas Indian Runners should be around 5lb. Some suggest the Penguin duck, now extinct, was the original, but this seems doubtful.

but the standardization came later.

The fact remains that the utility type were very much in evidence, being sold as Indian Runners even though they had been crossed. Some bore little resemblance to the proper type of vertical duck with its unusual walk.

UTILITY Runner
Laid 335 eggs
in one year.

OPPOSITE
An advertisement from the Feathered World in the 1920s, showing Reginald Appleyard as the leading breeder. His logo represents the ideal Runner.

Reginald Appleyard,
IXWORTH,
U.D.C.
I.R.D.C.
U.I.R.D.C.
SUFFOLK.

Telegrams: APPLEYARD, IXWORTH
Station: THURSTON (L.N.E.R.) 4 miles

Breeder, Exhibitor and Exporter of **Champion White Indian Runners, Fawn, and Fawn and White Runners, both Exhibition and Utility.**

1924 Wins include:
Dairy Show—1st, 3rd, 3rd and V.H.C. *Palace*—1st, 1st, 2nd, 3rd, 3rd, Res. and Reserve Palace Medal Best Runner. *Olympia*—1st, 1st, 2nd, 3rd, Reserve, V.H.C. *Royal*—1st, 2nd. *Tenbury*—1st, 1st, 2nd, 3rd. *Leominster*—1st, 2nd, 3rd. *Helston*—1st, Spl., 3rd, Res. *Tottenham N.U.P.S.*—1st, Res. and Certificate for Best Runner. *Hull*—1st, 1st. *Hertford*—1st, Spl. *N.U.P.S., London*—1st. *Norwich*—1st, 2nd, 2nd, 3rd.

Test Successes with White Runners.
1st "National" 1921-22.
  Pen 38, 5 Ducks Laying. 1006 1st Grade Eggs, 44 weeks.
1st "National" *Daily Mail* Test 1922-23.
  Pen 12, 5 Ducks Laying, 1151 1st Grade Eggs, 48 weeks, win Silver Medal.
1st "National" 2-year Single Duck Test.
  No. 189, "The Champion," officially 2-year Tested White Runner, laid 285 1st year, 246 2nd year, making 541 1st Grade in 2 years.
1st Stapleford Test 1924.   2nd National Test 1924.
  3rd National Test 1924.

THE PROVED STRAIN OF WINTER LAYERS.
Mated Pens, Trios, Ducks, Drakes and winners generally for Sale.
EGGS IN SEASON. LIST & PHOTOS FREE. STATE REQUIREMENTS.

A BREEDER'S VIEWS*

Both from a fancier's point of view and from that of the utility man the Indian Runner is ideal, although it is not easy to breed a really outstanding Runner, perfect in colour and type.

The Runner is without a doubt the Game Fowl of the duck world—with its clean cut outline, short hard feathering, elegant carriage and graceful movements.

On the other hand, when bred for eggs, not forgetting type, it is certainly the Leghorn of the duck family—a really wonderful layer.

It is said that over a century ago, Runners were brought by a sea captain into a port in Cumberland as a gift for his farmer friends; they were believed to come from Malaya and China. The captain's friends bred from them and also used them for crossing with other ducks. It was found that they were great layers and a big demand sprang up for these " Penguin Drakes," so called because of their upright carriage. Unfortunately this led to trouble, for any old drake or duck with a slightly upright carriage was sold as a Runner.

No Whites were exhibited in the early days; it is fairly certain that they were non-existent at that date.

---

* Reginald Appleyard, a famous breeder.

INDIAN RUNNER & BALI DUCKS 21

Runner Ducks:
Top: White
***Bottom*:** Fawn

Depicting the type bred by Reginald Appleyard a famous water fowl breeder.

One thing certain is that the Indian Runner is not a breed made by the fancier; improved if you wish, but not a manufactured breed. It is still possible to import Fawn Runners which can win in the very hottest company.

BREEDING AS A BASIS FOR UNDERSTANDING

The best way to understand the points of a Runner is to breed them ! It is not easy to breed a real " top notcher " in any of the colours, and we have Fawns, Fawn and Whites, Whites, Blacks, Chocolates and other colours, possibly too many because they tend to get mixed together.

The Blacks were bred by mating a brilliantly coloured Black East Indian duck with a Fawn drake and vice versa. The mating gives Chocolates as well as Blacks; also of course, many poorly coloured birds of nondescript colours—muddy Fawns and poorly coloured Blacks. One also loses type, but this is put right by careful in-breeding.

The beginner always thinks of the Runner as a very long, slim bird and such are easy to breed. One must, however, breed a bird of medium size, not too big and coarse but, on the other hand, not a weedy little bird with no stamina. Start to breed a medium-sized bird with short, crisp, glossy feathering, clean cut in outline, compact in stern, and with nice length of body and a natural carriage of at least 60-70 degrees. By this is meant carriage maintained when in a pen in the open and on the alert, not in a show pen.

INDIAN RUNNER & BALI DUCKS

In a good Runner the body must be perfectly round—*not flat on the back or chest*. Get either of these faults in a bird and it is certainly not a good Runner either from an exhibition or a utility point of view, for such a bird will lack lung and heart space.

The eye must be bright and prominent and placed high in the skull, as near to the top of the head as possible. The skull must be lean and rather flat on top. A coarse, round skull and the eye low spoils the appearance in that it gives what we term a " clock faced " Runner.

The bill should be of medium length, strong, and fits imperceptibly into the skull; if you get a weak bill which is concave on the top line you do away with the wedge shaped skull and bill and get what is termed a " dished bill " a very bad fault in a Runner. On the other hand, if the bill is too strong and inclined to be convex on the top line you have what is called a "Roman nosed " bill.

The great essential is to breed for the happy medium; in other words, one wants a clean line from the top of the bill to a point over and behind the eye.

The Runner is a bird of movement and without good sound legs and feet it cannot move with what is termed a perfect "outrun"; going away or coming towards you the bird must not waddle, as does an ordinary duck, it must go with a perfect outrun, head up, stern down, upright carriage, perfect balance, and without the slightest trace of a waddle.

The legs of a true Runner are set on to the body much further back than in the other breeds. I like legs of nice length with good sized feet and strong thighs. Given these and you get movement and a real Indian Runner.

As a layer, I consider the **Fawn and White** to be about the best, closely followed by the White. The former usually beat the Whites because they are not quite so temperamental and nervous—but much depends on strain.

INDIAN RUNNER & BALI DUCKS

On free range the Runner takes a lot of beating, ranging far and wide and picking up much of its living at certain seasons of the year in the form of natural foods. On the other hand, they seem to do very well without swimming water and are quite fertile on land.

INDIAN RUNNER & BALI DUCKS

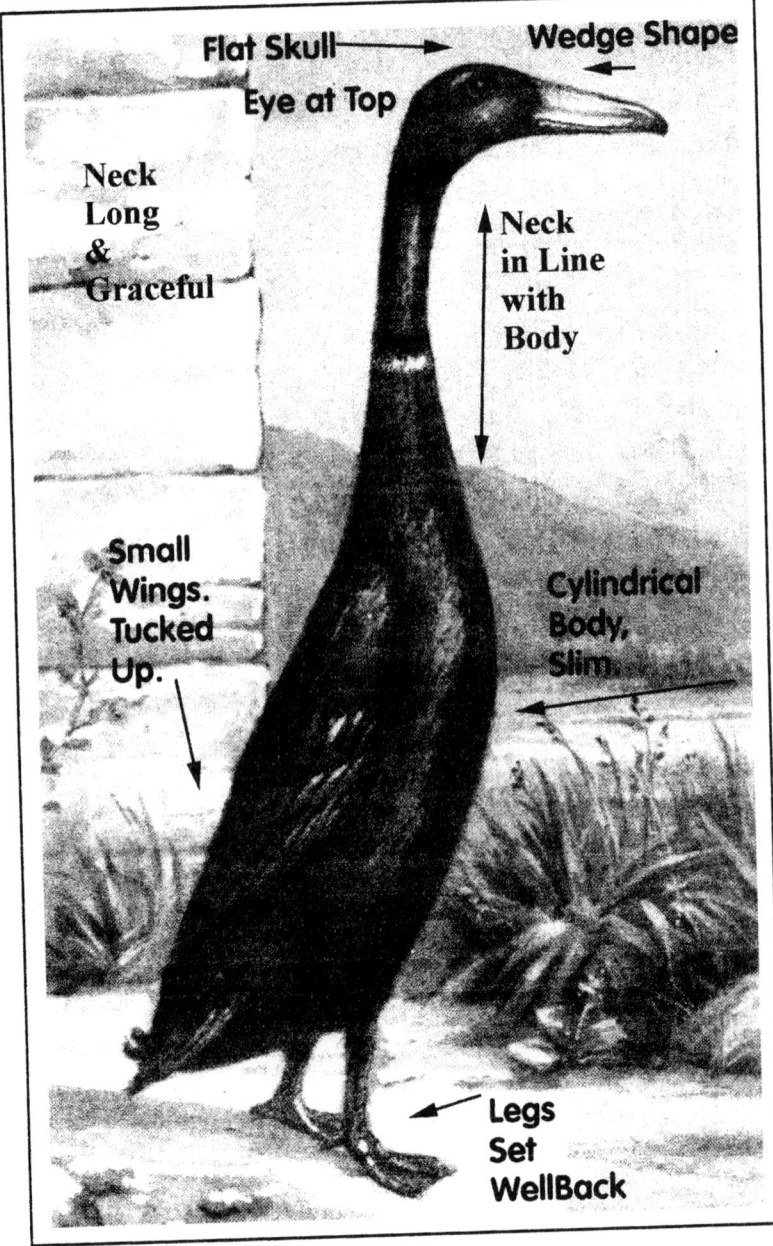

3

THE RUNNER DUCK STANDARDS

> **OPPOSITE**
> Ideal Features of Runner Duck

STANDARD OF PERFECTION

The *Standard of Perfection* laid down by the **Indian Runner Duck Club,** when functioning, was as follows:—

General Characteristics—The Indian Runner, as compared with the larger domesticated varieties, is a small, hard-feathered duck of a very upright carriage, and active habits.

Its body appears elongated and somewhat cylindrical, the legs being set on very far back.

NECK.—Fine, long and graceful, and when the bird is on the move or standing at attention is carried almost in a line with the body, the head being carried high and slightly forward.

BODY.—Slim, lengthy and rounded. From the shoulder points to the hip joints it should be nearly a cylinder. It is a little flattened across the shoulders, however. At the lower extremity the body sweeps round gradually to the tail, which is neat and compact and carried almost in the body line or horizontally, but should not be elevated or tilted upwards. The position or carriage of the tail varies with the attitude of the duck, but habitually upturned sterns and tails are objectionable.

The stern appears short compared with other breeds. The prominence of the abdomen and stern varies in ducks according to the season and the age of the bird, being fuller when in lay, but moderation in this point is to be observed. The large pendulous abdomen, often accompanied by long stern, and the other extreme, the " cut away " abdomen and stern are to be avoided.

INDIAN RUNNER & BALI DUCKS

WINGS.—Small in proportion to the size of the bird, are carried tightly packed to the body and well tucked up. The tips of the long flights of opposite wings cross each other over the rump of the bird, more particularly when standing at attention. At the upper extremity the body contracts to form a funnel-shaped process, which again gradually and imperceptibly, without obvious junction, merges into the neck proper, which is long and slender until it again expands slightly to be fitted neatly into the head.

The lower or thickest portion of this funnel-shaped process or neck expansion should be reckoned as part of the body, and the distance from the top of the skull to where the neck proper joins the thick part of the funnel should be about one-third of the total length of the bird.

The thinnest part of the neck is approximately that part where, in drakes, the dark bronze portion of the head and upper neck joins the lower or fawn portion of the neck proper. The neck should be neatly fitted to the head so as to maintain a clean, racy appearance. The muscular part of the neck should be well marked, rounded, and stand out from the windpipe and gullet, the extreme hardness of feather helping to accentuate this. The long axis of the body should pass through the centre of the funnel-shaped process (or neck-expansion), but as the head is carried high and slightly forward the greater part of the head will lie in front of this axis.

HEAD.—With the bill, wedge-shaped, lean and racy looking. The skull is flat on the top and the orbit, or eye-socket, is set extremely high so that its upper margin seems almost to project above the line of the skull.

EYE.—Full, bright, and alert.

BILL.—At its base should be fitted imperceptibly into the skull.

LEGS.—Set far back on the body so as to allow of an upright carriage. The thighs are longer than in most ducks and are strong and muscular; the shanks are short and **the feet** neat and supple.

TOTAL LENGTH.—Drakes, 26 to 32 inches(65 -80cm). Ducks, 24 to 28 inches (60 - 70cm).

WEIGHT.—Drakes, 3.50 to 5 lbs. Ducks, 3 to 4-50 lbs.

Note: (2.25k = 5lb)

Faults

Above the Standard weights and measurements, short squat bodies or oval-shaped bodies, long stern, wry tail, slipped wing or any deformity, domed skull with central position of eye, dished and weak bills, Roman nose, under-curved bills, thick necks and short necks, swan or curved necks, the neck-expansion set on too far back on the body causing a chesty appearance with a hollow behind, flattened backs or bodies, legs set on too far forward causing poor carriage; waddling or rolling action, a natural carriage in any duck below the minimum 40 degrees.

The USA standards also mention square shoulders, deep keels, low set eyes.

4

COLOUR DESCRIPTIONS

COLOURS—A BONE OF CONTENTION

Unfortunately, the duck fancier has allowed slight variations in colour to be developed as separate varieties. This discourages newcomers to the Fancy because the whole process of recognition becomes very difficult.

For example, what is a Fawn colour—is it a deep gingery colour, or is it lemon or is it brown or some other shade. It is left to the judge to decide and he is not given much guidance from the British or American Standards. If it is so difficult why bring in a different colour (ASP) called Buff?

In its standards the Poultry Club shows a coloured illustration for a Fawn & White which is certainly *not* fawn, but is a colour which might be termed a Grey-Chocolate & White (this is not a suggestion for a new variety!!).

The USA standards show a Fawn & White, male and female, which appear an acceptable shade, but the Pencilled Male illustrated is virtually the same as the male Fawn & White; it is slightly darker, but there is not much difference.

We are attempting to breed birds which come within accepted parameters of a colour shade and there are bound to be variations, so why produce slightly different birds and then call them something else?

COLOURS

Although it seems that the Fawn and White was the original colour, since that time many new colours have been added. A summary is shown below:

    1. Fawn

    2. Fawn & White

    3. White

    4. Black

    5. Chocolate

    6. Trout

    7. Pencilled

    8. Grey

    9. Mallard

    10. Cumberland Blue.

    11. Buff

    12.. Various Off Colours

THE FAWN

In Fawn **ducks** the general coloration should appear an almost uniform warm ginger-fawn throughout, with no marked variation in shade in the different parts of the body. The wing-bars are of the same colour but a few shades darker. The body exhibits a slightly mottled or speckled appearance. The bill is black, feet and shanks black or occasionally dark tan. The iris of the eye a golden brown colour.

When taken in hand the head and neck, also the lower part of the chest and abdomen, may appear a shade lighter than the rest of the body. Each small feather of the head and neck is lined with a one line of dark reddish-brown, giving the head and neck a ticked appearance. **The lower part** of the neck and the neck-expansion are a shade warmer in colour, each feather pencilled with warm reddish-brown.

In young ducks which have just completed the first adult moult there is often a rosy tinge on the lower neck-expansion, upper part of breast, and shoulders, but this soon fades away. The scapulars (or long pointed feathers on each side of the back covering the roots of the wings) a rich ginger-fawn, just a shade darker than the shoulder and back, with a wellmarked red-brown pencilling.

The body of the wing or bow is a shade lighter than the scapulars but darkening towards the wing-bars, the feathers pencilled as before.

Some ducks have a cream or light coloured narrow band in the wing-bar owing to the upper part of each feather being of a lighter or almost cream shade edged or laced again with the normal dark shade. The secondary flights are of a warm red-brown.

The primary flights are usually a shade lighter than the secondaries. The shade of colour darkens and is richer on the back and rump of the bird, the pencilling also being richer and more marked, but the ground colour becomes lighter and warmer towards the tail. The feathers of the tail are lighter in shade, each feather being pencilled.

The belly-colour is lighter than that of the upper parts of the body, being about the same shade of fawn as the head and neck, becoming a shade darker on the tail cushion, all feathers being pencilled as before.

FAWN DRAKES IN FULL PLUMAGE.—The head and upper part of neck are of a dark bronze colour with metallic sheen, which may show a faint green tinge. This dark portion should meet the colour of the lower part of the neck with a clean cut or the lower colour merge into it imperceptibly. The lower neck and neck-expansion are of a rich brown-red which is continued on to the breast and over the top of the shoulders and upwards to where it joins the head and neck colour, merging gradually on the back and breast into the body colour. Sometimes this brown-red or " claret" is absent, the French-grey of the lower chest and abdomen extending right up to the bronze of the upper neck. The lower chest, flanks and abdomen are of a French-grey, which is made up of a very minute and dense peppering of dark brown, almost black, dots on nearly white ground, giving a general grey effect without any show of white. This grey extends beyond the vent until it meets the dark, or almost black feathers of the cushion under the tail.

The scapulars (or long pointed feathers on each side of the back covering the roots of the wings) a rich ginger-fawn, just a shade darker than the shoulder and back, with a well-marked red-brown pencilling.

The body of the wing or bow is a shade lighter than the scapulars but darkening towards the wing-bars, which is fawn. The primary flights are brown, fairly dark in shade. The iris of the eye is often darker than in ducks. The bills vary from pure black to olive green mottled with black and with a black bean. Legs and feet may be black or darkish mottled with black.

When the drake is in eclipse he assumes the plumage is a dirty fawn or ash colour, thus approaching the colour of the duck.

INDIAN RUNNER & BALI DUCKS 37

THE
INDIAN RUNNER DUCK CLUB

Exists to encourage the breeding and exhibiting of the Pure Runner Duck, as imported from the East.

New members are invited to join.

Particulars from the Secretary—

FAULTS IN COLOURING OF FAWNS.

White anywhere; eye brows or eye stripes; light or cream wings (bows, coverts and flights) in the duck; blue or green wing-bars; orange or yellow bills, feet, or legs.

NOTE: The problem of deciding what is meant by Fawn is always present and is examined on page 32. In poultry showing for fowl and ducks the colour question causes many discussions and arguments. We all think we know a colour, but then discover there are many interpretations used by fanciers and judges. The Oxford Dictionary gives a definition *as a light yellowish brown colour.*
On this basis a lemon colour is not a fawn and neither is a deep brown colour.

It would follow therefore than any departure from the yellowish light brown should constitute a fault.

INDIAN RUNNER & BALI DUCKS

F. & W. Runner Drake, "DUKE OF WELLINGTON," is a son of " Lady Knights" the World's Record Breaker at the New Zealand Laying Test, 1920-21. She laid 333 eggs in 335 days. Grandaughters of this Drake are winners at the National Test this year, over all breeds, gaining the " Daily Mail" Cup, "Feathered World" Breed Cup, L. & N.E. Rly. Cup and Gold Medal.

THE FAWN-AND-WHITE

The head should be adorned with a fawn cap and cheek markings, nearly the same colour as that of the fawn of the body in the case of ducks but of a dull, bronzy green shade in the drake. The cap is separated from the cheek markings by a projection from the white portion of the neck, extending up to and in most cases terminating in a narrow line more or less encircling the eye. The cap should come round the back of the skull with a clean sweep. There should be no " tails " to it. The cheek markings should not extend on to the neck.

The bill should be divided from the head markings by a narrow prolongation of the neck-white, from one-eighth to one-quarter of an inch in width extending or projecting from the white underneath the chin.

When the bird is young the bill is light orange-yellow, but soon green spots show themselves and gradually unite extending over the whole mandible so that it becomes entirely, or almost entirely, of a dull cucumber shade in the duck, and a greenish-yellow in the drake. This is what should be bred for, though a darker shade is permissible.

The neck should be pure white to about where the funnel-shaped expansion begins, and should meet the body fawn with a clean cut.

Fawn & White Runner Drake
This colour is on the dark side.

White Runner Duck

From paintings by Kurt Zander.

Plate Showing Fawn & White Colour

The fawn should be uniform throughout the body—a soft warm or ginger fawn—the shade, however, will depend to some extent upon the amount of fading and bleaching caused by exposure and sunlight, which rapidly destroys the colour.

Colour uniform to the skin, the undercolour not of a different kind.

This description applies to the fawn over the **whole surface** plumage except the rump and tail of the drake, including the under surface of the tail, which are of a similar hue to his head.

When closely examined the coloured body feathers of the drake show a soft warm ground colour slightly peppered with a rather warmer shade. As only the outer edge of the feather is visible the colour seems solid and more ruddy than in the duck.

The duck should have much the same shade of fawn as the whole Fawn duck. The fawn and the white should **meet on the breast** with an even cut about half-way between the point of the breast bone and the legs.

The base of the neck, upper part of the wings, back and tail should be as nearly as possible the same colour as the fawn of the breast, and from the fawn of the back an irregular branch is thrown off on either side and extends downwards on the thighs to, or nearly to, the hough. The white of the breast extends downwards between the legs to beyond the vent and may overlap the thighs in part.

PENCILLED*

A medium fawn colour, but the duck has distinct pencilling. The head is a bronze green for the drake and fawn for the duck.

The neck is white for 66.6 per cent at the top and then the fawn appears in a distinct section.

GREY VARIETY*

The colours follow the lines of the wild drake and duck, the Mallard. This has been described as follows:

Male: Head & Neck brilliant metallic green, with a white ring at the baseBack & Upper Breast rich chestnut; Lower Breast and under parts greyish white, pencilled with fine dark lines; Tail four centre feathers curled upwards; Wing Bar blue, violet and green; Feet orange; Bill yellowish-green.

Female: Overall shades of grey and brown; white throat; Tail not curled.   Also see *Mallard*

BUFF*

Overall a deep buff or seal brown. This is similar to the British Fawn, except there is a clear indication that a positive buff is required.

---

* Recognized by the *American Standards (ASP)*

If the bird is coloured between the ribs and thigh it is termed " foul flanked." The primary and secondary flights should be pure white, as **also the** lower part of the wing-bow. The legs and feet should be orange-red.

THE WHITE

Should be a pure white. The iris of the eye is blue and the bill, legs and feet orange-yellow.

Stains or black streaks on the bills or mottled legs and webs are minor faults, for which a point is deducted.

THE BLACK.

Black Runners should be a solid black throughout, and have a metallic lustre like the Black East Indian. Bill, legs and web are also black or dark tan.

THE CHOCOLATE

This is an offshoot from the Blacks and should be a rich even chocolate throughout. The drakes on assuming adult plumage become darker than the ducks, but the ground-work is the same. The bill, as also the legs and webs, are black.

TROUT COLOURED

Drake similar to Mallard, but body tending to be silvery grey. The female is a silvery grey colour. Bill light olive and legs and webs dull orange.

THE MALLARD

Follows the colour of the wild duck. Drake has a green head and neck which is separated from the body by a white ring. Breast and back a dark burgundy and the under parts a blue grey. The wings are brown with a metallic blue wing band. The female is an overall medium brown with black markings.

CUMBERLAND BLUE

Blue overall with darker shades on each feather. Bill should be bluish-grey in the female and bluish-green in the male, whereas the webs and feet are orange.

SCALE OF STANDARD POINTS

Head, eyes, bill and neck (exclusive of lower neck expansion)...........	20
Body, shape, and general appearance of (including lower part of neck, legs and feet)...........	35
Carriage, and Action	30
Colour and Condition...........	15
	100

This follows similar lines to the Poultry Club.

EXHIBITION versus UTILITY

As far as we are aware no research has been done on the effects of slimming the body on egg-laying potential. The early specimens of Runner had rotund bodies which were full at the bottom and they were excellent layers. Is there a danger in producing these very slim birds of reducing the laying potential? We believe that this is a possibility, although this conjecture is based on experience with other layers and not factual information on Runner ducks. However, we should not forget the utility aspects of these fine breed.

INDIAN RUNNER & BALI DUCKS

JUDGING RUNNER DUCKS

Judging is generally done on the basis of comparison, viewing the features of one against the others in the class at the show. The features to examine are:

1. Shape

This should resemble a narrow bottle, at an almost vertical angle. The body should be slim and round and surmounted by a slender neck which bends very little, and a head which has a flat skull and wedge-like bill. The latter should be straight from the top of the head and be quite long.

Birds which are coarse, or have a flat back, or heavy shoulders or deep keel shouls be penalized.

2. Carriage

Vertical carriage is essential. Top specimens stand quite straight with little or no tendency to lean forward. The posture should be graceful and when this feature is present it is immediately apparent; no slouchers should be given prizes.

Faults include swan necks and feet which are in the wrong position.

3. Movement

Runner ducks do really *run* and certainly do not waddle like other ducks. The keen duck keepers believed in training their birds to become tame and to watch them run so that only those with the correct quick movement would be used for breeding. Ideally, at shows, the Runners should be seen on the run, but this is not always possible. If this is possible, exhibitors should be able to watch the birds being judged on the run.

Any sign of waddling should be penalized.

PENGUIN DUCK

DESCRIPTION

The Penguin duck was recognized by the Poultry Club in 1923 its sponsor being Mr Harold W. Andreae. The original Penguin has been known for many years, but the "new" breed sponsored were selected for laying qualities and had been modified. The breed is sometimes said to be similar to the Indian Runner, but a photograph in the "Poultry" Year Book for 1924 shows a duck rather like the Magpie duck, but a little more upright.

The colour is black with a white front, from the underside of the neck extending down the breast to the legs. The wings are black with white primaries. The weight is 7 to 9lb. for the drake, with the duck 1lb. lighter.

The bill should be slate colour and the legs as dark as possible. A full Standard was given in the Poultry Club Standards for 1930 and this refers to the body being as wide and long as possible with a broad back, which are not characteristics of the Runner duck.

The carriage is expected to be almost upright, but the modern Standard stipulates that the Penguin is inclined to be like the Pekin.

Sadly the breed is now very scarce and may even be extinct although it is always dangerous to assume this fact, because, quite likely, there are flocks running around in some remote spot.

DUCK MANAGEMENT

GENERAL RULES

DUCK MANAGEMENT

NATURE OF DUCKS

Ducks are hardy creatures and can thrive in conditions that would deter normal breeds of fowl such as Leghorns or Orpingtons. Moreover, many of them lay extremely well, although some people never seem to take to duck eggs, even though they are quite nutritious and delicious.

The fact remains that given the market or need ducks can be quite profitable. Moreover, the heavier breeds, such as the Aylesbury, Rouen or Pekin make excellent table birds and these are in constant demand in a gourmet type market.

1.CLASSIFICATION

Attempts have been made to classify ducks into different categories as follows:

LIGHT (5lb; 2.25k)

Abacot Ranger

Bali

Khaki Campbell (almost Medium)

Coaley Fawn

Magpie (almost Medium)

Runner

MEDIUM (8lb; 3.60k)

Buff Orpington

Blue Swedish

---

* *The American Standard of Perfection* suggests this approach.

DUCK MANAGEMENT

Cayuga

Crested

Huttegem

Orpington

Penguin

Saxony

Stanbridge White

Welsh Harlequin

HEAVY (8lb +; 3.60k+)

Aylesbury

Baldwin

Blue Termonde

Pekin

Muscovy

Rouen

Rouen Clare

Saxony

Silver Appleyard (Standard size)

BANTAM SIZE (40oz; approx 1k)

Call or Decoy

East Indian (recognized by APA)

Mallard

Silver Appleyard Bantam Duck

DUCK MANAGEMENT

CLASSIFICATION

CHARACTERISTICS

1. The "upper weights" given are only a guide and there is a little overlap. It will be seen that the two top layers — Campbell and Runners appear in the Light class.

The dual purpose breeds (laying and table) come into the Medium range and the table breeds are the Heavy weights.

2. EGG NUMBERS, SIZE AND COLOUR

Eggs for most ducks are quite large and may approach 3 oz. (85g). The medium layers will produce, say, 100 eggs per annum and others will lay 150 and more; the top layers such as Khaki Campbells and Runners will lay in excess of 300 eggs per annum. The objectives will determine which breed to keep.

On colour some breeds produce white eggs whereas others will have green or blue-green. Generally speaking the white eggs are to be preferred; duck eggs are viewd with suspicion anyway so they are more likely to be accepted if a conventional white. Runner ducks have white eggs.

3. FLESH COLOUR

In some countries the white fleshed bird is preferred. In any event a dark coloured duck will have dark stubs when plucked and this detracts from its appearance. These are factors to be cosidered when deciding which breed to keep.

4. OTHER FACTORS

The space available and whether a pond and grass are available are factors to consider. Some breeds like to forage; others prefer to have a pond; other breeds such as Indian Runners and Campbells do not seem to mind the absence of a pond.

SPECIAL NOTE

Bibliography

For more details on the management and breeding of ducks, readers are referred to:

Domesticated Ducks & Geese, J Batty

PLEASURE FROM DUCKS

Duck keeping can provide hours of enjoyment and relaxation as well as eggs and table birds. In the Spring the ducklings provide new interest and enable the hobby or business to be of greater variety to the dedicated breeder.

The fact remains that the ducks must be selected on the basis of what is available in terms of pasture, orchard or paddock. Do not overcrowd or try to keep ducks on unsuitable ground; in fact, a marshy pasture or a neglected orchard, quite unsuitable for any other livestock, could be suitable for profitable duck-keeping. Remember that ducks do not scratch up the garden so they do not disturb growing plants.

If a pond is available all the better because then any type of duck can be kept and they can be given all the natural rements. Besides the profit angle there is the aesthetic appeal; rippling water, beautiful ducks swimming around — and even the most soberly-plumed ducks are a joy to watch on the water. However, Runners do not seem to need a pond.

DUCK MANAGEMENT

A Home-Made Pond

Ducks like Runners, Muscovies and Campbells do not appear to miss not having a pond, but a small bath of some kind is desirable.

CROSS BREEDS

When keeping ducks for commercial purposes some cross one pure breed with another to exploit to the full some particular aspect or to bring in more vigour to the stock, Aylesbury X Pekin is a common cross because this aids in breeding because the Pekin drake is more active. This gives more rapid growth to the ducklings.

Other possible matings are :

1. White Campbell X Mallard

If the male is a Mallard the offspring have light coloured flesh, but darker with a male Khaki Campbell.

2. Rouen X Aylesbury

Produce good tasting birds with a gamey flavour.

Rouen X Pekin are similar.

3. Runner X Pekin

This cross gives more eggs than the Rouen and also produces a small table bird.

For normal purposes it is most unwise to cross the breeds because once crossed they cannot be shown and the many years of work of staning the breed are lost. In fact, it is wise to ensure they do not cross-breed by keeping the breeding pens separate.

DUCK MANAGEMENT

ACCOMMODATION

Ducks need dry accommodation which is well ventilated. Various types of floor can be used and success in keeping the floor dry has been achieved by the use of wire (twilweld) or slats, using special wood that will withstand dampness (duly tarred or creosoted).

Pens need not be elaborate, but make sure the ducks are safe from foxes and other vermin such as rats.

Ducklings should not be reared on wooden floors because this may affect their walking and can cause lameness.

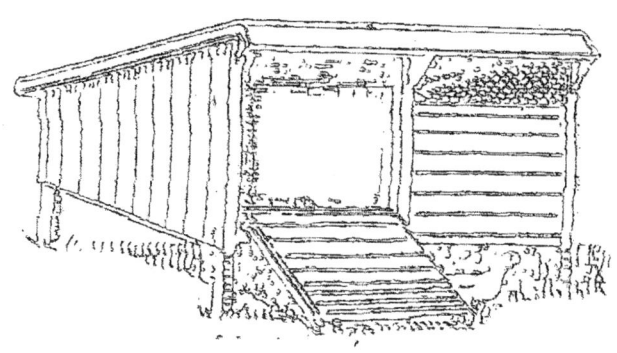

Duck Shed with Ramp
This can be closed when the ducks have gone in.
They may have to be kept indoors for a week and then driven back until they are trained to go in themselves.

FEEDING

Ducks are very hardy and in suitable surroundings will find a great deal of their own natural food such as grass, grubs and insects. However, this should be supplemented with corn soaked in water and, when laying, with layers' pellets or mash. Fattening ducklings require special feeding of high protein food.

The provision of water is vital for most breeds, although the Runners and Muscovies do not seem bothered. and, if a pond is not supplied, there should be a water container deep enough for the ducks to immerse their heads, thus allowing them to dabble.

Grit should also be available both naturally and in a trough so that the neccessary calcium is consumed. Duck eggs are large so a considerable supply of soluble grit is taken.

Possible Diets

There are differences of opinion on the diet to give. Some give poultry layers' pellets and corn in a water trough. Others vary the diet using different grains and additives.

Ducks lay 200 eggs or more so there should be no neglect of the food requirements. Give sufficient food to allow them to fill themselves in, say, 20 minutes and then move the troughs. Corn in water will allow them to top up in the evenings.

DUCK MANAGEMENT

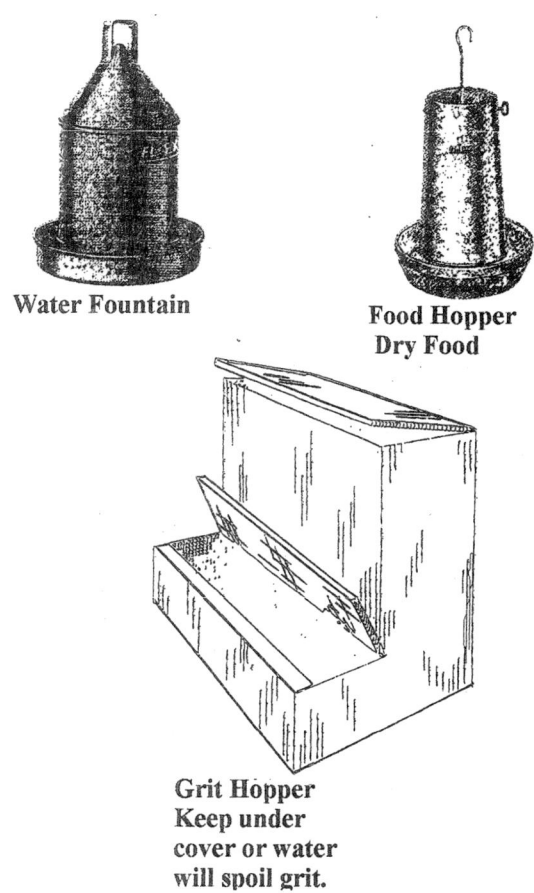

Water Fountain

Food Hopper
Dry Food

Grit Hopper
Keep under
cover or water
will spoil grit.

Possible Basic Utensils

An open trough might be a useful addition for holding water and throwing in wheat for feeding, but must be cleaned out regularly and fresh, clean water added.

A very experienced duck-keeper who believes in natural foods gives his views as follows:

1. Give soft food in the mornings.

This should be equal parts of barley meal, bran, sharps, boiled vegetables with cod liver oil or fish meal added. This may be varied according to the ingredients available, but should be balanced so that sufficient protein is present—about 20 per cent.

2. Wheat put in a Water Trough.

This allows the ducks to take the soaked wheat when desired and does not result in waste food being taken by rats and birds.

3. Adequate Greens

Normally a plentiful supply of greens will be available outside in paddock, garden or orchard, but in winter raw swede turnips, put through a mincer will be a substitute.

Some ducks, including Runners and Campbells, are excellent foragers so at most times of the year can pick up worms, snails, and other tit bits, as well as greens in the form of grass and weeds.

DUCK MANAGEMENT

BREEDING

For the active breeds there should be no problem obtaining fertile eggs and hatching them under a broody hen or incubator.

Some ducks will hatch their own, but this stops egg production and therefore a surrogate mother may be preferred.

The pen should be made up of an active drake and ducks with a known record for laying well into the Autumn. If for the table then earlier on, and when rearing the ducklings note and ring the rapid developers for they could be future breeders.

For show purposes a "one to one ratio" may be advisable, especially when dealing with the top heavy breeds Aylesbury and Rouen. With these, with their size and deep keels, there may be some difficulty anyway, so do not try to put too many ducks in a pen.

Usually 3 ducks to a drake is the correct ratio with more when the ducks are a very active breed. Unlike chickens (where cocks will fight) there can be two or more drakes running in a pen with a number of ducks. However, for careful monitoring of the results separate pens with a drake and ducks is better; trap nesting might be preferred when a strain is being built up.

---

*For details see *Natural Incubation & Rearing* or *Artificial Incubation & Rearing* both written by Dr J Batty

DUCK MANAGEMENT

When catching them up drive into a shed and then close the door. If necessary use a large net to trap each duck required. Pass a hand under the body and hold close to your body until quite firmly held.

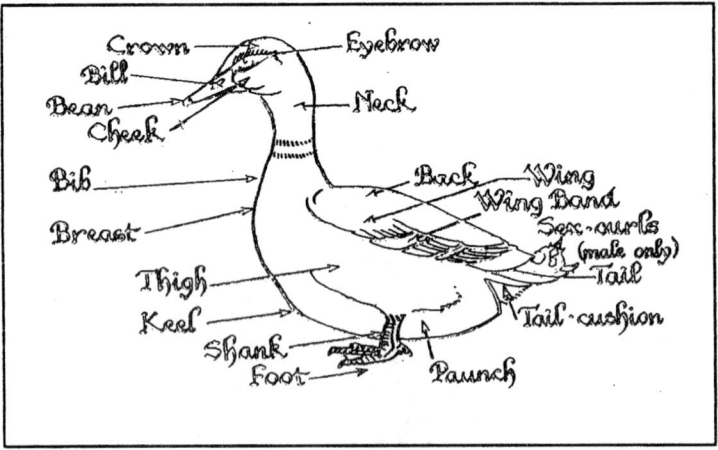

Features of the Duck

Note the Runner has an upright stance and must be very slim,

DUCK MANAGEMENT 63

NOTES ON BREEDING RESULTS & EGG PRODUCTION

....

....

....

INDEX

Accommodation 57

Breeds 1-49, 50, 51, 52, 53.

Classification 53
Cross Breeding 56

General Rules 49

Feeding 58-60

Ponds 54,55

Utensils 59,60